Printed and distributed in the United States of America

Edition 2
21.3.23.1M

https://www.bookpopmedia.com

Important Note:

Although BookPop Media and the author of this book strive for the highest quality in the resources offered here, unless otherwise noted, BookPop Media and the author are not responsible for the validity or accuracy of the material presented in this work. Materials in this book are thought to be accurate, but no warranties, expressed or implied, are made. BookPop Media and the author of this book specifically disclaim any warranty of merchantability or warranty of fitness for a particular purpose. BookPop Media, its staff, and the author of this book will not be held liable for any direct, consequential, or other damages resulting from the negligence of BookPop Media or its agents. The use of this book constitutes understanding and acceptance of these provisions.

Always consult a trained mental health professional before making any decision regarding treatment choice or changes in your treatment. Never discontinue treatment or medication without first consulting your physician, clinician, or therapist. If you feel like you want to harm yourself or others, please consult a mental health professional or other health care professional IMMEDIATELY. Self-help resources such as this book are not meant to nor cannot replace the specialized training and professional judgment of a health care or mental health care professional.

The information provided in this book is designed to support, not replace, the relationship that exists between a patient/site visitor and his/her existing physician. The contents of this book should be considered for entertainment purposes only.

That is why it's important to remember that this book alone is not meant to and cannot replace the specialized training and professional judgment of a health care or mental health care professional. You should consult your doctor, therapist, or other healthcare professional if you have any questions about your personal health or psychological needs and before implementing or making changes in any treatment or habits.

Table of Contents:

I dedicate this book to everyone who has suffered anxiety or depression, in any degree, because of their relationship with their smartphone. In a social media connected world, we are all in it together, but can still feel lonely. Our days may start with our smartphones and end with our smartphones, but with balance and mindfulness, we need not suffer from them. I hope this book will help you along your path to coming to terms with (and hopefully defeating) any smartphone-related problems you may have. We have never met, but I believe in you.

— B.R Wang

Mobile Mindfulness:

A Guide to Identifying and Resolving Your Smartphone Addiction and Anxiety

B.R Wang

CHAPTER ONE

Introduction to Smartphone Mindfulness

THE COVID-19 SMARTPHONE EFFECT

The year 2020 was an unforgettable year for those who experienced the COVID-19 pandemic and spent much of our time at home. Less likely to go outside, people's smartphone and internet usage skyrocketed as our devices became our primary news and entertainment sources. According to the research, *Preventing problematic internet use during the COVID-19 pandemic: Consensus guidance*[1], information, and communications technologies were the "savior" for distributing information about the pandemic and served a critical role in keeping society intact. Our smartphones, computers, and other devices helped us work, study, entertain, stay in touch with friends and family, and other possible functionality. The only major problem with staying at home so much and having such

[1] Király, Orsolya et al. "Preventing problematic internet use during the COVID-19 pandemic: Consensus guidance." Comprehensive psychiatry vol. 100 (2020): 152180. doi:10.1016/j.comppsych.2020.152180 (https://www.ncbi.nlm.nih.gov/pmc/articles/PMC7215166/)

convenient and powerful technology is the potential for addiction and psychological issues substantially increased. The research paper mentioned above states: "While considered healthy when pursued in moderation and for meaningful purposes, excessive engagement in specific online activities such as gambling, viewing of pornography, video gaming, social media use, shopping may lead to severe problems and elevate the risk of disordered or addictive use." When I was writing this book, life was returning to normal, but many of us still depend on our smartphones and tablets more than ever!

60 DAYS A YEAR

Before the COVID-19 pandemic, twenty percent of smartphone users were on their phones for approximately three to four hours a day, according to the *RescueTime Blog's Screen Time Stats of 2019*[2]. **These stats imply people could spend about 60 full days of their life on their smartphones every year!** That's a lot of time, and life is too short to waste it on things that don't always matter! The RescueTime post covers many exciting smartphone research topics, and we will cover many similar topics in this book. If you have time or want to review some data they reported, check out their resources linked at the end of this chapter or at the end of the book if you are reading this in an E-Book.

For us, smartphones bring many comforts to our life. With our technologically impressive smartphones, we can

[2] RescueTime Blog's Screen Time Stats of 2019 (https://blog.rescuetime.com/screen-time-stats-2018)

check the news or access anything we want in less than a minute. Consequently, the smartphone has also brought many downsides to society. Some of us have become too attached to our phones and are thus too dependent on them. Many of us cannot leave our phones at home, and should we find that we left our phones somewhere, we could nearly experience a panic attack!

If you're one of those people, you might have a smartphone addiction or consider yourself possibly over attached to your phone. Either way, hooked or not, you want to establish some adjustments to your habits to live a happier and healthier life. Any efforts you invest will pay off if you remain consistent!

If you've come here searching for a cure to the constant anxiety brought on by your phone, you're at the right place. This book is going to help you organize peace with your phone!

Although not yet considered a formal psychological disorder, smartphone addiction is dangerous enough to affect productivity. It can cause depression and become one of the biggest obstacles to your success and happiness.

MY STORY AND WHY THIS BOOK EXISTS

In my late teens and early adulthood, I went through smartphone addiction on and off. For me, smartphone games, news feeds, and constant need for digital attention were my weaknesses. Every 5 minutes at work, I would sneak a look at my phone to see if anything new was happening. If a notification came, I would stop whatever I was doing to check it and react. I'd continuously compose photos and post them to multiple social media sites and await the incoming flood of comments and likes. Over time, parts of my offline life indirectly suffered, and I was not even aware of it. Small problems were compounding and they were subtly correlated with my increased smartphone and internet usage.

As I gradually became more anxious about the world overall, I tried to figure out why. After a lot of soul searching, I realized my internet relationship was the root of all my troubles! I knew that stopping the use of my smartphone was impractical; it is such a useful tool! I recognized that deleting all of my social media apps was too much. For some people I knew from school, social media was the only practical way to stay in touch with them. In conclusion, it was on me to learn how to control myself and not let my smartphone negatively affect me. From there, I began my research on smartphone addiction.

If your phone is causing more trouble than it's worth, it is time to rethink your relation with your phone. However, the solution is NOT to get rid of your phone entirely and downgrade to a 90s "dumb phone" with no social media, internet, or apps. This book's goal is to promote you to find balance and middle ground in your relationship with your smartphone. For that reason, I wrote this work.

This book aims to support you with:

- Determining whether you're possibly addicted to your smartphone.
- Showing potential root causes.
- Finding peace with your smartphone addiction if you have it.
- Noticing how you can use minimalism to enjoy

your smartphone without getting too attached to it.

I wrote this book as a guide to identifying the causes of and conquering smartphone addiction. Some of us, including myself, dose off when reading large chunks of paragraphs, so each chapter is broken down into bite-sized pieces for easier readability and digestion. We'll work together to help you discover the power your smartphone has over you and how you can take back control! You don't need a smartphone or tablet to live a happy and healthy life!

CHAPTER TWO

Nomophobia and Anxiety

WHAT IS NOMOPHOBIA?

Regardless of whether it is a minor or a substantial occurrence, Nomophobia can happen to anyone, and being aware of the anxiety is the first step to coming to terms with it. Nomophobia derives its name from the phrase "NO MObile PHOne PhoBIA." Characterized by a feeling of panic or uneasiness among smartphone users because of a lack of having their mobile phone nearby, Nomophobia affects people of all ages. Smartphone usage habits are the most significant factor in whether someone will feel anxiety about being away from their smartphone. It is primarily not the same as a typical phobia like arachnophobia (the fear of spiders), but more of an anxiety disorder. Still, it fits our smartphone addiction terms, showing that smartphone addiction is more than just an informal anxiety disorder.

According to the *2019 Connectivity and Mobile Trends*

Survey of Deloitte[3], every American checks their phones more than 40 times in a day. These results imply that Nomophobia needs to be considered a serious condition, and people who may suffer because of their smartphones should know the symptoms. To determine whether you or someone you know may have Nomophobia, here are some of the signs and manifestations of smartphone addiction:

A SECRET LOVE-HATE RELATIONSHIP WITH YOUR SMARTPHONE

Here's the most extreme Nomophobia situation: you're subconsciously in love with your smartphone, and you can't comfortably go through your day without it. You live every moment of your life with your phone by your side and even leaving it in another room makes you wonder whether you are missing out on anything. If you spend some time watching TV or cooking dinner, you still keep your phone nearby just in case you need it for something or that it requires you. If you don't have it, you miss it!

You feel lonely and disconnected without having your phone. Maybe the surrounding people in your actual life are boring or don't share your interests. Coincidently, the individuals on your social media accounts seem more exciting. As a result, you can't stop checking in on them.

At the same time, you despise social media and news sites. You're tired of everything bugging you and the shallowness you see everywhere. You crave something "real" or with greater purpose and substance, but you find

[3] Deloitte Connectivity and Mobile Trends Survey (https://tinyurl.com/y2wt4x2x)

yourself coming back to your smartphone every day anyway. You realize something is wrong with how you are managing your device or that you are using it too much. It's a great tool, but also a double-edged sword, and you know you need it regardless.

SOCIAL DETACHMENT

Your attachment to your smartphone can lead to social detachment. When virtual reality feels more exciting than the real world, it may indicate that you have Nomophobia. Even when you're attending a social gathering, people may often see you using your smartphone instead of taking part in the moment. Remember that talking with your friends or family is also a helpful way to release stress. When you're spending your precious time on your phone instead of interacting with your buddies and family, it may mean that you're addicted to your phone.

RELATIONSHIP STRAINS

When you're living in virtual reality, it can take a toll on your real-world relationships. You are mentally absent from your loved ones' lives, which leads to a weaker bond with your friends and family. Sometimes, phones connect us better to people who are farther away. Paying attention to how your constant smartphone usage negatively affects your relationships is essential.

Be it the continuous checking of Instagram, stock market prices, work emails, pornography, or your other favorite app of choice, the underlying denominator is

that you are balancing your cell phone life with your relationships. Whether it is unbalanced attention or your smartphone interfering with moments that should be intimate is where you should draw the line as to whether you may be addicted. A good question to ask yourself is if the present moment with your friends or family would have been higher quality if you hadn't spent the time on your phone.

CONSTANT PHONE CHECKING

You might face the following subconscious scenario daily or even multiple times a day:

Are there any new notifications? What's new? Why haven't I heard back from them yet? Why isn't there any news on the latest news topic yet?

Okay, there are no new notifications on my lock screen, but maybe something else is posted. Let me go look at the other app now and see if anything is new there...

Does finding yourself continually checking your apps to see if you have something new to read sound familiar? Not only this, but you often go through the statuses of your friends on social media, even for those you don't find interesting. Satisfying that curiosity cures the boredom, that is true. But at what cost do you keep yourself entertained?

TIME ADDS UP

Have you ever counted or tracked how much time you spend on your phone? If you're hooked to your phone, you're likely wasting time while scrolling down the newsfeed. You may also forget, ignore, or procrastinate doing many everyday tasks around the house, even those you promised to do for someone else.

If you want to gain awareness of your smartphone usage, you need to try this:

- Whenever you handle your phone, start the stopwatch or note what time it is.
- When you stop using your phone, write down how much time you've spent on your phone.
- Estimate the usage amount for the entire day. If you're spending more than an hour without doing something meaningful on your phone, you're possibly addicted to your phone.
- Count even a "quick check" of your notifications or "phone pickups." Potential addiction has two measurable signs:
    - The frequency of manually checking for new or updated information
    - The time spent endlessly/mindlessly browsing a news feed

FEELING DISCOMFORT OR PANIC WHEN YOU DON'T HAVE YOUR PHONE NEARBY

You have a class, meeting, or an event where you can't bring your phone in or that it has to be entirely off. When you are unable to check your phone, you can't help but wonder what you are missing out on or if something important is going on. When the meeting completes and you can recheck your phone, you feel relieved, even if no new notifications came when your cell phone was away or off.

NEWSFEED ANXIETY AND DEPRESSION

Do you ever read the latest posts and feel like crap afterward because of some bad news or toxic online communities? Do you ever feel you're just endlessly scrolling through your newsfeed in search of something that will make you feel happy or that there is a missing piece of the puzzle in your life that will inspire you out there somewhere? Your dopamine levels may be in trouble! We will cover why this happens and how to get through this numbness in our *The Dopamine Download* chapter.

REAL LIFE IMBALANCES

When you're addicted to your phone, there's sometimes a thin line between virtual reality and real life. As a result, it affects your actual life. It is a common occurrence not to find anything other than your phone or the people in it interesting. From reading books to socializing with your friends, your phone appears to be the only source of your entertainment and information.

Consequently, your phone is your only comfort-zone in this world. When you feel sad, you go to your phone. When you want to entertain yourself, you go to your phone. And soon, you're stuck with your phone, and your real-life exists little to none. You don't want to hit the gym or vent in-person to your friend. If the conditions above sound familiar to you, you're possibly addicted to your phone.

PHANTOM NOTIFICATION VIBRATIONS

It's possible that you sometimes feel the vibration of notification in your pocket without getting one. Smartphone addiction can cause your body to "feel" an alert even when you aren't actually receiving one.

WARNING SIGNS ARE THE BOTTOM LINE

The signs and manifestations mentioned above can be a warning sign for you. If some of these symptoms are happening in your life, take a step back and reflect. Is your smartphone negatively affecting your feelings, productivity, or relationships in any way? Keep in mind that addiction is a spectrum, not binary. That means that one person can be more or less addicted to something than their friend. In the upcoming chapters, we will help you build a healthy relationship with your phone and use it wisely.

CHAPTER THREE

Signs of Smartphone Addiction

SMARTPHONES, ADDICTIONS, AND OUR BRAINS

Addiction is a concept most of us are familiar with; however, it's not consistently obvious what real addiction is versus harmless frequent enjoyment. With addiction, a person feels a compulsive dependency on something. Sometimes, what the individual is hooked on is inherently harmless, but if it develops into an addiction, then it will be apparent because of the inevitable problems that arise. What makes smartphone addiction harmful? The frequency and amount of consumption!

If you look around, you'll see people with different addictions related to anything or any behavior. Most of these people usually are addicted to consuming alcohol, marijuana, caffeine, video games, tobacco, or other hobbies and substances. The question isn't what, but why?

Why do such people adhere to the consumption of these substances? Though useful in many fields of life, why do these consumables cause a negative impression and influence when they are the center of addiction?

Let's consider water. Water is vital for us; we literally cannot live without it. But if you drink too much of it, you will die. Essential items have positive effects, but these benefits stop if the frequency or amount of usage goes unchecked or out of balance with what we naturally need. With overconsumption, simple things become harmful.

When people think of addiction, they think of drugs, alcohol, or video games first, but I'm here to tell you that smartphones can mess with our brains in similar ways.

In the later chapters, we'll cover:

- Signs of smartphone addiction
- Whether you are addicted
- Causes of anxiety associated with smartphone addiction
- And a lot more!

Smartphones are gadgets that allow us thousands of benefits: contact with our loved ones, availability of universal information, and plenty of other functions! What's important is that you peek into your life and daily routine. Are you subconsciously using your smartphone excessively? Are you genuinely connected with the genuine people around you? Are you trying to find a kind of shelter from your problems on your smartphone? Is your health facing unnoticed risks?

You may have already tried testing and analyzing if you or someone you know is possibly a smartphone addict.

Manifold are the factors that can cause mobile phone

addiction among people. Do any of these below items apply to you?

SMARTPHONE EASE AND PORTABILITY

Have you ever questioned why you don't carry a laptop with you all the time, despite that it can also fulfill your needs? Most of the time, a traditional laptop can still store and operate the same apps and games that you regularly use.

The reason is that smartphones are portable and make the apps and programs you use even more accessible on demand. Being small and easily carried, smartphones are always in the pocket of your pants or purse. They are a lot easier to pull out and check for information than carrying around an entire laptop. The portability is great but also allows for a more straightforward pathway to addiction.

STRESS, DEPRESSION, ANXIETY, AND LONELINESS

The use of smartphones can be a source of pleasure you can carry with you. Smartphones can take us away from the mess we feel in our lives. They prove to be a valuable escape from all the troublesome conditions and boredom we suffer from daily.

When a student in the scientific study, *Internet Addiction: A Brief Summary of Research and Practice*[4], was asked the

[4] Cash, Hilarie et al. "Internet Addiction: A Brief Summary of Research and Practice." Current psychiatry reviews vol. 8,4 (2012): 292-298. doi:10.2174/157340012803520513 (https://www.ncbi.nlm.nih.gov/pmc/

question about his addiction to his smartphone, he replied,

> "I feel technology has brought so much joy into my life. No other activity relaxes me or stimulates me like technology. However, when depression hits, I tend to use technology as a way of retreating and isolating."

The student's statement made me wonder whether depressed and stressed people are more vulnerable to smartphone addiction than those who are happy and already content with their lives. Let's investigate!

A DESIRE TO SOCIALLY CONNECT

Social media allows us to stay connected with people who we care about, even if we only moderately care about them. But with smartphones, this connection becomes artificial. As discussed above, the world of virtual friends and following celebrities subconsciously gives us feelings of concern about them all the time. We always want to know:

- What our friends are up to
- What the celebrities are doing
- The news that's happening in places where we don't even live!

The thing is, most of the time, Hollywood stars and celebrities don't even know you exist, yet you follow their daily moves! Such virtual social connections are also the

---

articles/PMC3480687)

cause of addictions to smartphones among many people.

FEAR OF MISSING OUT (FOMO) ON NEWS AND SOCIAL MEDIA

When we think about our friends, relatives, and celebrities who may enjoy their lives more than us, we subconsciously dive into the quest to identify what they do that's making them so happy. People want to see what is causing peoples' ways of life to be so exciting and entertaining!

Fear of Missing Out is a kind of feeling where you don't want to miss out on any news because you feel a compulsive need to know the latest stories and get updates about it in real-time. You habitually load up your favorite social media pages and groups every day to check if anything new is happening or announced.

In the BBC News article, *FOMO: How the Fear of Missing Out drives social media 'addiction'*[5], they tested social media addiction among several students. One student felt unproductive without his phone. Another student reported that he developed a jealousy of those who adhered to their phones. It shows that with smartphone addiction, one develops a constant curiosity and an inner compulsion to check their social media accounts, newsfeed, and other apps to know what is happening worldwide. Endless newsfeeds can be a vector for smartphone addiction. Especially with online news, this can trigger anxiety and various other feelings across the internet world.

---

[5] FOMO: How the Fear of Missing Out drives social media "addiction" (https://www.bbc.com/news/technology-39129228)

NOT ALL INFORMATION IS WORTH ACKNOWLEDGING

For a moment, consider whether everything you hear or read on social media immediately affects you. Every time you check your social media or news apps, your mind gets a dose of information that has very little to do with your practical life. For example, your newsfeed may contain a mix of something like these examples:

- A music concert is happening in China next week, but you live in the USA.
- A celebrity you heard of broke up with another star.
- A top YouTube streamer dislikes the rumors of an upcoming video game and is angry about it.
- A football team player has a leg injury, but you don't want football.
- A new DIY recipe to make picture frames out of old Amazon boxes is available to learn.
- Speculation about an approaching election is happening in a news article, and several people are debating about it.
- A stranger online posted a photo of something mildly interesting they saw in their backyard.

Did reading all of those bullet points take some energy to visualize in your mind? If you have an endless stream of mismatched, unrelated posts of various topics, your brain is going to have to continually work to process and imagine the contents of what pops up on your screen. Over time, that stimulation will cost you a lot of mental energy.

If you're getting tired, can you drop what doesn't really affect you or matter? News and entertainment you see online is not always 100% factual or based on reality. Most of the time, the story we hear is deliberately tweaked to spread emotion about a particular issue. If news were written just as facts, it wouldn't be interesting enough for many people to click.

CHAPTER FOUR

News: Authenticity and Reliability in the 24/7 News Cycle

AN OCEAN OF UPDATES

A never-stopping cycle of information and news surrounds us, be it national news, world news, politics, social, economic trends, tech announcements, and other topics. This continuous supply of communication makes it almost impossible for some of us to stay away from it. If you are spending a lot of time or following too many topics, carefully examine which news is relevant for you and which is not. Even if entertaining or essential at a global scale, too much information can cause you unneeded stress and mental fatigue.

THE PROBLEM OF FAKE NEWS

Nowadays, regardless of political leanings, the news's authenticity is a big question on everyone's minds. The clickbait headlines and news headlines are so carefully designed that it becomes challenging to scrutinize what is truth, what is an opinion, and what they tweaked to gather clicks. Most of the time, the news articles get written with an agenda, typically aimed at getting individuals to click or subscribe to the platform. Moreover, the best way to attract more people is to create a sensation in the information.

SENSATIONALIST NEWS

You'll often see thumbnails on news and social media sites are highlighted with bright colors, bold text, and catchy phrases. These expressions or sentences are short and precise but still convey a sense of urgency, scandalous, or dramatic content. You also get a lot of news through your smartphone that, after independently verifying, ends up only being rumors, a sliver of the whole truth, or the "most interesting" possible outcomes. In reality, the news that provokes excitement and vigor among the readers is not neutral. So, you must keep your eyes open for what is the actual information and what is just sensationalist clickbait.

NEWS OR OPINION?

Almost half of the individuals scrolling down their newsfeeds cannot identify whether they are reading a piece of news or an opinion piece. According to the American Press Institute article, *Confusion about what's news and what's opinion is a big problem, but journalists can help solve it*[6], only 43 percent of people using social media can identify news versus opinion. About 73 percent of people believe that opinion is an unnecessary point when highlighted as news.

Here are some reasons for such a flood of misleading stories, opinion injected news, and sensationalism around us.

POLITICAL INTERESTS

Many famous political personalities and organizations are notorious for giving disparaging comments on their counterparts. The reason is simple; to create height in their character and show faults in their colleagues' characteristics. Such activity promotes discord among already aggressive followers and pushes people farther away from the middle of the political spectrum.

RIVALRY BASED PORTRAYAL

Political smearing is not unique only to the political figures. Celebrities with large amounts of followers are also engaged in negative portrayals of their rivals. Such ridiculous commentary is not worth your time. Like the example of the music concert happening in China, this doesn't affect you, so it's not worth paying attention to it! Stars have political views too, and whether they are left or right, they have influence.

---

[6] Confusion about what's news and what's opinion is a big problem, but journalists can help solve it (https://www.americanpressinstitute.org/publications/confusion-about-whats-news-and-whats-opinion-is-a-big-problem-but-journalists-can-help-solve-it/)

BEHIND IT ALL: EARNING REVENUE

As discussed earlier, companies create sensationalism and fake news to attract massive viewership and traffic. The more the subscriptions and visits an article gets, the more advertising revenue these sites earn. Not all stories are fake news, but keep in mind that the motivation to make money is always there, even for supposedly "independent" news organizations and YouTube channels. Publishers will invariably need money to operate, and reports that sell or gain clicks will always be more profitable to publish than plain, unbiased, straightforward news.

WHAT DOES THE RESEARCH SAY ABOUT NEWS CRAVINGS?

Many platforms have researched why we crave to have a constant supply of knowledge. A research article, *Smartphone usage and increased risk of mobile phone addiction: A concurrent study*[7] within the International Journal of Pharmaceutical Investigation, suggests that most participants knew about some risks of smartphones, but developed dependent behavior with them anyway. From checking notifications to connecting with our friends, we stare at the screen of our smartphones. Untracked habitual peeking into our phones can become mentally harmful and

[7] Parasuraman S, Sam AT, Yee SWK, Chuon BLC, Ren LY. Smartphone usage and increased risk of mobile phone addiction: A concurrent study. Int J Pharm Investig. 2017;7(3):125-131. doi:10.4103/jphi.JPHI_56_17 (https://www.ncbi.nlm.nih.gov/pmc/articles/PMC5680647)

can cause depression and anxiety.

DEVELOPING HUMAN INTEREST

Different issues and happenings in the world increase our interest in humanity and injustices. Many of the people interested in such topics find themselves emotionally stimulated by the causes of society. Social activists and philanthropists contribute their part in aiding the underserved and underprivileged, inspiring many others to do the same. Following the latest information on social issues can be great, but be sure that the constant news is not negatively affecting your mood too often.

FORMATION OF OPINIONS

If we indulge in the streams of different opinions and points of view, we can also develop our own. We become confident and profound in our knowledge about various topics; thus, we can contribute to our society's innovative ideas. Some people scour over forums and message boards every day to find like-minded opinions and encourage others to learn their perspectives. The problem is that these message boards often become echo chambers where people with the majority point of view "rule" the board. Whether intentionally or by accident, differing opinions get buried into inactive sections of the website. The "echo chamber" happens because the most popular topics gain the highest visibility on social media websites because of how the algorithms build the newsfeed. This situation can evolve into extreme cases of confirmation bias about various issues such as politics and history. If you hold an unpopular opinion, you may even feel like a minority because of how the "majority" always wins presence on the social media feed. These social media situations are very complicated and vary from community to community. Nevertheless, trending content and blow-up of popular opinion potentially leads to psychological problems for all users.

NOT EVERY RUMOR MATTERS

You don't need to know everything. Is your notification screen full of the next iPhone's announcements when you are not even planning to purchase it? What about a rumor about two celebrities who might be dating? Is it worth your attention? What do you have to do with the stories when you know that those are rumors actually and have nothing directly to do with your life? It is essential to consider whether you should remove that app's notification privileges in such conditions, so you aren't bothered by them anymore.

MOST WORLD NEWS WON'T AFFECT YOU

How excited are you by reading that Xi Zing has won the Magnificence Award for her stunning acting? Of course, it is not that exciting or relevant to you unless your name is Xi Zing!

The news about the harvest of cocoa in Africa will not affect you when you live in the USA. Similarly, a fire happening in another country will not approach you here in your house. It would be best if you remembered that not everything in this world would directly influence your life.

Keeping up with world news is vital for global awareness, but consider how much of your life you are sacrificing to keep up to date with everything as it is happening. Beware not to become emotionally invested in headlines and news that do not directly affect your life. If you read every news feed on every website posted ever, you'd never be able to stop reading! Think of how much time and energy it would cost you to consume all of that content, and what benefit would you have from it besides dinner conversation topics?

NOT EVERYONE NEEDS FIRSTHAND INFORMATION

If your friend is addicted to knowing every news story firsthand, is it necessary for you to do the same? The latest essential news will reach your awareness shortly after the event by word of mouth from acquaintances and family; you need not actively pursue the newest information. You should know that not everything is necessarily relevant for you and that by letting the news come to you from friends instead of seeking it out, you can save so much time. Unless there is a crisis happening nearby and you need the local news for your safety or security, getting the information live as it comes out probably isn't fundamental.

HEADLINE STRESS DISORDER

The news that we read daily in our newsfeed is not always pleasant and entertaining. Sometimes, it may be a piece of bad news: heartbreaking, threatening, disappointing, or full of stress. Continuously staying in touch with this kind of 24/7 news cycle can cause certain psychological and physical strains on your body.

According to the Healthline article, *Headline Stress Disorder: When Breaking News Is Bad for Your Health*[8], suffering from "Headline Stress Disorder" causes someone to feel continuous anxiety, tiredness, and stress to the point where it can even affect their real-world relationships.

Now that you've finished this chapter, you can be self-aware of the smartphone addiction problems and how news affects us. You can discern whether a website is trying to get you to click on their link/video and identify likely biases in news sources and celebrities. You've also learned how too much "clickbait" can cause mental stresses and other anxiety-related troubles. In the next chapter, we will cover all about push notifications and how they can also affect your well-being.

---

[8] Headline Stress Disorder: When Breaking News Is Bad for Your Health (2019): (https://www.healthline.com/health-news/headline-stress-disorder-when-breaking-news-is-bad-for-health)

CHAPTER FIVE

Why Am I Getting This Notification?

THE RELATIONSHIP BETWEEN FOMO AND NOTIFICATIONS

If you are suffering from the Fear Of Missing Out (FOMO) on the news, you might find it challenging to discover whether your smartphone's notifications are valuable and worth keeping. You may fear that you'll miss important pieces of information if you turn them off.

Whenever you hear a ringtone or get a vibration on your smartphone, you know that it's an incoming call or a push notification. If your smartphone has many apps, there's a higher chance that your smartphone's lock screen gets filled with lots of messages every day.

Push notifications are means by which the developers of the apps installed on your smartphone remain in touch with you. But do you really need these interruptions?

HOW MANY PUSH NOTIFICATIONS DO YOU GET?

The statistics given by the *Business of Apps Push Notification Statistics (2019)*[9], shows that according to the Accengage Software Company, an average cell phone user in the US gets 46 pop-up application messages every day. It also tells that the users' opt-in rate for these push notifications is higher in Android than in iOS. The more important question is how many smartphone users find all of these notifications useful?

PLEASE PAY ATTENTION: NOTIFICATIONS WANT YOUR ATTENTION

The primary purpose of these push notifications is to engage the users in an app and interact. You know very well that you are the customers of the apps you have installed on your smartphone, whether the app was free. Either way, using the app leads to the developer earning money from you in some shape or form.

Many of the apps work in collaboration with business-oriented purposes and goals. They earn revenue from these platforms by advertising goods and services or collecting usage data. That aside, it means you can unconsciously become a part of their marketing and earnings-based strategy. Unfortunately, getting lots of notifications also

---

[9] Business of Apps Push Notification Statistics (2019) (https://www.businessofapps.com/marketplace/push-notifications/research/push-notifications-statistics)

costs your brain energy. With every message you read, you sell a bit of your life to give the application a chance to earn some money.

STIMULATION BY DESIGN

By design, push notifications make money off of the smartphone users, both directly or indirectly. It is also vital to discover why some of us easily fall into this trap. The answer is simple: push notifications play with our minds, and they know how to trigger us.

Two kinds of factors stimulate the needs or urges in our mind to do anything: external and internal factors.

When you have been starving for a day, your stomach will signal you to eat something; this is internal triggering. When you are full and have no space in your belly for anything else, it signals you that you're full. This feeling is how internal and external factors work.

Can you imagine what you'd do if both circumstances happened simultaneously; you'd be starving and full at the same time, whereas both eating and not eating would be painful. The sensations alone would tire you out! All you could think about is food, which is the strategy of app notifications in a nutshell.

App notifications combine both of the external and internal stimulation factors and put them together in front of you.

For example, let's say you are planning to go to the grocery store soon. Suddenly, you see a notification from an e-commerce app that gives you the chance to order the exact same grocery items you need, but with no delivery charges! It's likely that you'll opt for the free delivery from the app and stay home too. Now the next time you want to get groceries, you might check that app and see if there's still a promotion going on. Maybe the next time the app pushes you a notification, you'll consider ordering from the app, even if there's a small fee the next time around. The thing is that once you place an order in the app once, you're more likely to place another order in the app later, especially since:

- You already installed the app.
- The app already has your shipping and billing information.
- You have used the app or website and it has already "proven" itself and with you subconsciously.

The developers want their users to think of their apps at every opportunity so you'll always consider using the app when relevant or possible.

Relevance is also a point that such apps don't ignore. They know that such offers will only work out when the advertisement is suitable for the user. Targeted ads and push notifications reinforce this process of building a relationship with you, the user.

NOTIFICATIONS FROM YOUR FRIENDS

Not everyone around the world uses their smartphones for informative, educational, or professional purposes. Most people only use their phones for entertainment. They engage in different activities through their smartphones, such as gaming, chatting, photos, video sharing, and a lot more.

Nowadays, memes are something prevalent all over social media. If you are a meme lover, you probably already know what they are and what they do. For those of us who aren't quite familiar, memes are usually pictures, quotes, puns, or anything else that fulfills the function of satirizing or making a statement about something. These memes are concise and typically share an idea funnily or sarcastically.

When someone finds something humorous, unusual, or related to their friends on social media, they tag their friends to share the post. In the same way, when your friends tag you in a meme or reach out to you on social media, a notification pops up on your screen. Out of curiosity or addiction, you'll interact with that notification!

Then what happens? You'll forget about it within an hour! Therefore, these notifications are not worth your attention all the time! And remember, the messages add up!

WHAT IF THE NOTIFICATIONS WERE HUMAN BEINGS?

Suppose you wake up on Monday morning and twelve people are standing beside your bed, smiling at you and holding different signs in their hands for you to look at. They all are trying to get ahead of each other to claim your undivided attention. What would your reaction be?

Most of us would be annoyed or stressed out by people bombarding them with information the very moment the morning alarm goes off! The strange thing is, we've already become accustomed to and made a habit of checking our lock screens right as we wake up. Some individuals even check their phones before acknowledging their spouse!

If you find the hypothetical situation of people crowding around your bed in the morning acceptable, let's consider this alternative scenario.

Don't you hate it when someone rings your doorbell, and you go to the front door to see who is your unexpected visitor and what they want? It interrupts whatever you were in the middle of doing, and most of the time, it isn't something important. When notifications arrive to your smartphone, the same effect of your ringing doorbell happens to your brain!

How do you find it? It can be irrelevant or annoying 99% of the time if you have many apps pestering you for your attention.

So, here, understand that all this mess is irritating, and consider making up a plan to cope with your phone demanding your recognition at its own will. To control this problem, you have to figure out what apps are essential for your device and what is not. It would be best if you analyzed what is worth your thought and your time. Your attention is a valuable asset; don't give it away for nothing!

Who would you let stand by your bed in the morning to talk to you? How many people are a crowd?

YOU CAN DEAL WITH IT!

Every app has an option to mute or disable its notifications. You can either mute those notifications and go through them at your leisure or disable them if you don't want to check them even in your free time.

You can also opt for "Do Not Disturb Mode" or such

alternative features on your phone to schedule periods of allowed interruptions. Remember, most smartphones will enable you to set exceptions from individual contacts so they can reach you in emergencies even if you've turned on your Do Not Disturb mode.

According to a *VWO Engage report on Why Users Mute Notifications*[10], around 52% of the users say they find push notifications annoying and irritating. Almost 49% claim that such notifications distract them from what they are doing.

In conclusion, try to make notifications work for you by checking them at your convenience. Don't let the notifications control your life!

[10] Why Users Mute Push Notifications (https://buildfire.com/why-users-mute-push-notifications/)

CHAPTER SIX

A Physical Dependency

THE PHYSICAL RELATIONSHIP WITH OUR SMARTPHONES

Though smartphones are not physically part of our bodies, we have a strong attachment to them. A lot of us have an uncomfortable feeling if we are without our phones for a while. Some of us even feel some anxiety if we leave our phone in another room!

In *Dependency on Smartphone Use and Its Association with Anxiety in Korea*[11], Public Health Reports researched the connection between time spent on smartphones and anxiety. They found that as people use their phones more often, they are more likely to become dependent. Smartphone dependency led to an increase in overall stress. This research affirms that although compulsive

[11] Lee, Kyung Eun et al. "Dependency on Smartphone Use and Its Association with Anxiety in Korea." Public health reports (Washington, D.C. : 1974) vol. 131,3 (2016): 411-9. doi:10.1177/003335491613100307 (https://www.ncbi.nlm.nih.gov/pmc/articles/PMC4869088/)

smartphone usage is viewed as a pathological condition, it is undoubtedly increasing unnecessary dependence. Ultimately, this damages the well-being of addicted smartphone users. Increased utilization of smartphones identifies with emotional health issues, for example, intensifying discouragement and anxiety. Another study, *Smartphone Use and Physical Activity among College Students in Health Science-Related Majors in the United States and Thailand*[12] by the International Journal of Environmental Research and Public Health, says that being over-engaged with mobile phone usage is making our lifestyle sedentary.

Smartphones are providing almost all you can want at your doorsteps. Despite that smartphones are portable and can be carried anywhere, they eliminate most needs to leave the house. If someone wanted to, they could order all food, supplies, and other goods online for delivery and never leave their home! This habit can lead to negative impacts on the physical and mental well-being of both adults and kids. If a person isn't careful, they can kiss an active lifestyle goodbye!

OFFLOADING THE FEATURES

Here's a list of everyday actions most of us do through our smartphones. Check and see how dependent you are on your phone for these daily activities:

[12] Penglee, Nattika et al. "Smartphone Use and Physical Activity among College Students in Health Science-Related Majors in the United States and Thailand." International journal of environmental research and public health vol. 16,8 1315. 12 Apr. 2019, doi:10.3390/ijerph16081315 (https://www.ncbi.nlm.nih.gov/pmc/articles/PMC6517887/)

- Checking the date and time
- Managing To-do Lists
- Shopping
- Paying Bills
- Watching Movies
- Sports Scores
- Gaming
- Photography
- Homework
- News
- Checking your health-related stats
- Reading books
- Planning and doing travel
- Social Media

This list is not very long, but there are many other functions that your smartphone lets you do with its help. The bottom line is that these actions fall into two general categories: productivity or entertainment. If you open your phone to perform one of these activities, it is effortless to follow up and do one of the other actions too.

The question is whether it is really important to do all the things discussed above through your smartphone? Are there alternative methods to accomplish these actions? The answer is yes. People had been performing all the tasks, as mentioned earlier, before smartphones. But the complication is that we've opted for a sedentary life by handling all these matters through only one platform, our smartphones. It's great, but it's not!

The problem is that when we go to check on something important or relevant, it opens the gateway into checking other non-essential apps too. If every time you intended to look at your calendar on your phone, you inadvertently checked your Facebook too, perhaps you should consider these alternatives:

Leave your phone at a dedicated spot in your home if you aren't using it. You can choose to use real clocks and calendars throughout your house if you find you frequently need to refer to them.

- Use diaries, sticky notes, or a planner book to manage your daily routine.
- Travel to the library or read printed magazines.
- Go outside as much as possible for your life if it is safe and plausible to do so. For example: Do your shopping in a brick and mortar store instead of ordering everything online. A lot of places will price match Amazon.
- Spend quality time in public and accumulate real-world experience.
- Visit or call a doctor or medical institute for your health-related issues.

Some above items may seem old-fashioned. They might feel foreign if you haven't gone outside very much in a while, for whatever reason. Going back to pre-smartphone technology and methods may not be as efficient either, but doing tasks the slow way can help you stop checking on your smartphone too often. Go on an adventure for the sake of helping you become less dependent on your smartphone! Triggers that cause you to check your smartphone are never without their alternatives; the only thing you have to do is discover them. Know the difference between love, convenience, and addiction; notice what's truly beneficial to you and what's causing a feeling of hollowness.

Don't be a slave to your smartphone when you have the chance to live free!

CHAPTER SEVEN

The Dopamine Download

MY FIRST SMARTPHONE

I remember the time when I got my first smartphone. It was like having a small magical box in my hand. There were many things which I could do with it that I couldn't do before, even with my old Palm Pilot, iPod, and graphing calculator combined!

Even today, whenever I first use a new phone, there is a little ting of excitement in the back of my mind, much like opening an unexpected present on Christmas.

As discussed in the previous chapter, our smartphones have tons of functions that replace many other hardware pieces. However, there is a line between the usefulness of the phone and getting addicted to it.

ALL ABOUT DOPAMINE CYCLES

Dopamine is a neurotransmitter used by your brain to transfer messages between the nerve cells in your body. Its production and regulation within the brain are related to attention, sleep, motivation, pleasure, learning, and voluntary and involuntary actions.

Unregulated dopamine levels play an important role in anxiety, stress, and depression too. When we are addicted to drugs or a smartphone, our dopamine levels increase. That means the more the drugs or smartphone used, the happier and satisfied you feel, at least at first.

Your body has a natural reward system that gets satisfied with the increased dopamine heights, particularly at the beginning of an addiction. Thus, your addiction builds satisfaction in your natural reward system, leaving you feeling satisfied. However, this system slowly and gradually restricts itself after repeated usage of the particular things you are addicted to.

Repeated and long term use of anything increasing your dopamine levels also increases the threshold needed for peace of mind and joy. This means that over time you will have to do more and more to get the same amount of happiness as before.

This building of tolerance applies to smartphone users too! If you keep letting it stimulate you, you'll need to continually seek new dopamine triggers related to your smartphone. This dependency means that you must check on more apps, more social media sites, and cycle through checking them more often. If you've skimmed your apps and there were no new news articles or social media updates, you'll become disappointed or anxious from the lack of dopamine. With no dopamine when you need it, you'll wait a little and then check through your apps again to find something that will give you a good feeling. The cycle will continue over and over. This leads to how smartphone behavior has a connection to gambling!

THE CONNECTION TO GAMBLING

Smartphones provide many gambling-like opportunities in the virtual world. Many laws in various countries restrict traditional gambling in the sense of risking real-world cash to win real-world money. Where regulation falls short is in other modern gaming business models, including "free to play, but pay to win" types of games. Even though a game may be free to download, gamers will pay with their time, money, or both to feel satisfied. There are many pathways to smartphone addiction, but gaming can stand out as a more accessible route. Dopamine is at the center of it all!

Once addicted to their smartphone, users behave in ways similar to compulsive gamblers, such as:

- They have no or very little control over their impulse to use their cell phone.
- Not recognizing the time they are spending on their phone. A person can waste lots of time without even realizing it.
- Gamblers know inside they are losing their game more than winning, but they keep playing anyway. Likewise, smartphone addicts use the phone without withdrawing.
- There is no break for them unless their phone needs to charge.
- Excuses to keep playing may include:
    - "Just one more round!"
    - "Until the end of this chapter!"
    - "Until I'm ranked higher!" or

- "Until I hit the highest level!"

We are living in the modern world where gaming is acceptable as long as it is in moderation. We must remember that our most precious asset is our time. If we waste too much of it, we will carry regrets later in life, just like a compulsive gambler.

CONSTANT REWARDS AND FEELING GOOD

Rewards systems built into our games and apps are one pathway leading us to addiction. The satisfaction of our excitement assists us in getting addicted to our smartphone as a gambler becomes used to gambling.

Our smartphone apps trigger the reward system in our brains because human beings are already wired to react to reward and punishment regardless of the medium. For example, Likes, Comments, and Subscribe notifications are developed in the apps for the same purpose.

Keep in mind that dopamine is the released chemical whenever we feel good. Remember when I mentioned the excitement of using a new cell phone and opening a Christmas present? When a person uses their phone, we experience the constant release of dopamine; the "feel good" chemical.

The cycle of using the phone and feeling positive continues until we become too used to always using our smartphone. The same process leads us to where the device becomes the primary source of the "positive vibes" chemistry in our brains. Thus, the addiction likelihood increases because little else can make us feel as happy as often.

WHAT DOES THE RESEARCH SAY?

According to the book *Treatment for Stimulant Use Disorders*[13], high levels of dopamine in someone's mind mostly improve temperament and increase body development. However, a lot of dopamine may deliver anxiety, crabbiness, forcefulness, and neurosis that may lead to schizophrenia (a mental disorder).

Another research article, *Mobile phone mania: Arising global threat in public health*[14], solidifies that mobile phone addiction is like any other drug addiction! With mobile phones, it stirs your dopamine level in the same way as if you were taking a drug! The only difference is that your intake of a drug would physically be inside your body. The mental effects between drugs and smartphone usage are comparable!

Another study, *Cell-Phone Addiction: A Review*[15], postulates that addiction entails abuse without control. Alterations in mood, tolerance, abstinence, and personal harm are all applicable here too.

---

[13] Center for Substance Abuse Treatment. Treatment for Stimulant Use Disorders. Rockville (MD): Substance Abuse and Mental Health Services Administration (US); 1999. (Treatment Improvement Protocol (TIP) Series, No. 33.) Chapter 2—How Stimulants Affect the Brain and Behavior. Available from: (https://www.ncbi.nlm.nih.gov/books/NBK64328/)

[14] Peraman, Ramalingam, and Subramani Parasuraman. "Mobile phone mania: Arising global threat in public health." Journal of natural science, biology, and medicine vol. 7,2 (2016): 198-200. doi:10.4103/0976-9668.184712 (https://www.ncbi.nlm.nih.gov/pmc/articles/PMC4934115/)

[15] De-Sola Gutiérrez, José et al. "Cell-Phone Addiction: A Review." Frontiers in psychiatry vol. 7 175. 24 Oct. 2016, doi:10.3389/fpsyt.2016.00175 (https://www.ncbi.nlm.nih.gov/pmc/articles/PMC5076301/)

It's clear that if you are doing anything exciting again and again, it will disrupt your natural dopamine levels. You will only be content when you use your smartphone as an addict. Eventually, your dopamine levels will not reach the threshold needed for happiness, even if you somehow used your smartphone 24-7. By this point, your body would have lost the ability to produce dopamine naturally. Your reward system will be stuck searching for something pleasurable. If that pleasure isn't achieved, you will feel anxious and depressed, even while holding your smartphone.

LOSING CONTROL IN THE SEARCH FOR DOPAMINE

Like other addictions, the main trap of smartphone addiction is losing control. In this stage, the user already went through their smartphone's initial excitement and the pleasure of using its features. They get stuck mindlessly and compulsively checking their phones and feeling anxious or subtly upset if there are no new updates available on their news feed. App after app, they hunt for something new until they find the next thing that triggers a dopamine release. This craving and an endless search for fresh content is the most significant symptom of smartphone addiction.

CHAPTER EIGHT

What About Productivity?

GIVE AND TAKE

You know that your smartphone calls for constant recognition from you. You also understand that most of the time, you give your smartphone attention whenever it requires it. Here's a question: Do you fulfill the demands of everything else in your life the same way you do for your smartphone?

Are you always checking your smartphone when you have a free moment? Are you going through your news feeds regardless of whether you are with friends, family, eating, at work, or even in the bathroom?

You might have an idea that smartphones are harmless at these times, and yes, that's true to an extent. However, consider the following:

- Does your productivity require your smartphone for everything?
- Has smartphone usage made you more productive?
- In what ways has a smartphone reduced your

productivity?

PRODUCTIVITY AND INTERRUPTIONS

Have you seen the people around you getting appreciated for their excellent performance at the office or their creativity? Have you noticed some individuals seem to always have expertise and experience at their fingertips? What makes them so skilled and prominent?

It's their productivity and time management.

Your productivity is how efficiently you work in a relative amount of time. It tells if you are fulfilling the demands of a particular job and furnishing the required output in the expected amount of time. It reflects how dedicated and concentrated you are at your work. We all have the same number of hours in our day, but how we enhance or limit ourselves is greatly influenced by our decisions about how we will spend our time.

Everyone has interruptions and complications in their life, but how we deal with them affects our capacity. Productive people know how to put unimportant distractions aside and concentrate on their tasks. Thus, your productivity depends on the management of distractions that you can control.

Check out the questions below and reflect:

- Do you sleep and wake up at the same time almost every day?
- Do you finish all of your work as scheduled without procrastinating often?
- If you're a student: Do you complete your

homework on time? Do you study as many hours as recommended by your academic program?

- Do you accomplish your office tasks on time?

How does your smartphone help or hurt you with the above situations? Everyone is unique, so you may be reading this and thinking, "No, my smartphone does more good than harm." Regardless of whether your smartphone affects your productivity, consider whether you can optimize your relationship with your smartphone to help your productivity.

HOW IS MY SMARTPHONE AFFECTING MY PRODUCTIVITY?

Imagine that while you are busy working on something, you get a smartphone notification. You grab your phone or glance at your smartwatch to see what's going on. Lots of different apps and notifications are on your lock screen, so you glance through them. None of them were important, so you put your phone back down again and continue working. Everything is fine, right?

Not quite! Even by just checking your lock screen, you were interrupted and distracted! Your mind had to break concentration and switch to the small, but different task of processing through any alerts and notifications on your lock screen. As a result, your brain had to "reboot and recall" to return to where it was the moment before the phone disturbed you. If you were working on something complicated, the time and energy it takes to resume where you were at can be even more consuming.

If you want to keep yourself productive, you must create

up an environment that ensures minimal interruptions and disturbances. The more you engage with the distractions, the more you'll delay completing your task or make it more challenging to complete the work in the same time.

OBSESSION WITH PHOTOGRAPHY AND DESIRE FOR INTERACTION

Do you think you'll be your most optimized self if you are obsessed with taking selfies and working out your next Instagram post?

You'll waste lots of time planning your useless commentary that most people only will not even remember after scrolling past it. Plus, how do the likes and comments help you besides a moderate ego boost? Wouldn't having more free hours or minutes in the day be better?

To stay focused and productive, you must think carefully about whether carrying on a high-frequency social media presence is worth the tradeoff.

APP CHAINING

We've loaded our smartphones with several applications; some are even interconnected or related to each other.

Perhaps you have three news apps, and you always check on all three of them every morning out of habit. Would your world be any worse off if you deleted one app?

DISTURBED SLEEP CYCLE

As you know, keeping your eyes glued to your smartphone screen late at night can harm your sleeping patterns. Late-night smartphone use could also affect your capacity to get appropriate rest. Being unable to sleep on time can lead to a loss of freshness. It can also negatively influence your productivity the next day.

WHAT DOES THE RESEARCH SAY?

In Addictive Behavior Reports, *Smartphone addiction, daily interruptions and self-reported productivity*[16], researchers sought to identify the interrelations between smartphone addiction, smartphone disturbances, and work-related productivity.

Their research reveals that the use of smartphones at workplaces decreases the production of employees. Moreover, it says that smartphone usage also negatively affects the leisurely time activities of those individuals. This research highlighted how people missed work because of poor health and how over-engagement with smartphones affected their tasks. The investigation revealed that addicts spend a considerable amount of free time on their smartphones. The users feel compelled by the notifications on their phones and continuously check for new content. The resulting habit of regularly checking their smartphone can grow into an addiction.

[16] Duke, Éilish, and Christian Montag. "Smartphone addiction, daily interruptions and self-reported productivity." Addictive behaviors reports vol. 6 90-95. 19 Jul. 2017, doi:10.1016/j.abrep.2017.07.002 (https://www.ncbi.nlm.nih.gov/pmc/articles/PMC5800562/)

Another research article, *Problematic Mobile Phone Use, Academic Procrastination and Academic Performance of College Students*[17], highlighted a correlation between Problematic Mobile Phone (PMP) usage and procrastination. The researchers found that the overuse of smartphones among students resulted in delays when submitting their assignments.

BOOST PRODUCTIVITY BY SETTING A TIME LIMIT

Before starting a task at work or home, set a deadline for when you'll intend to complete that task. Make yourself deadline-oriented, and it should steer you in the right direction. If you have trouble meeting your deadlines because of smartphone-related distractions, try asking a friend to hold you accountable. If you find that you are making your deadlines too easy on yourself, tighten up a bit. You should feel comfortable with your deadlines if you are minimally handling your smartphone. If you're feeling too much pressure and already stopped using your smartphone excessively, lighten your workload or stretch the deadline a little farther.

Don't forget to put your phone into the "Do Not Disturb" mode while you are actively working toward your goal!

If you are having trouble staying productive, try doing the *Pomodoro Technique*[18] of splitting your work into 25-minute intervals with brief breaks in between each focused

[17] Qaisar, Shahzada & Akhter, Nasreen & Masood, Afsheen & Rashid, Sumaira. (2017). Problematic Mobile Phone Use, Academic Procrastination and Academic Performance of College Students. (https://tinyurl.com/yykdvjjy)
[18] The Pomodoro Technique: (http://www.pomodorotechnique.com/)

work session.

CHAPTER NINE

The Reason for Smartphone Anxiety

SMARTPHONES SHOULD MAKE ME HAPPY, RIGHT?

You use your phone because it gives you pleasure. Right? So why do some people feel anxiety even though they are using it all day long? Besides FOMO on news and notification fatigue, here are some other sources of smartphone-based stress:

EXCESSIVE RELIANCE

We have molded ourselves as people who can't survive without having our smartphones in our hands. How many of us would comfortably drive to a new address using only verbal instructions and no GPS?

What happens when you see your battery is getting low or if your internet data limit is about to end? If you were on the way somewhere and your phone was disabled, how much panic would you have?

We have stopped relying on our skills, creative power, and intuition. Such dependence isn't harmful until it's gone, but it will cause stress when you no longer have your smartphone.

ANXIETY AND DISTURBING THE SLEEP CYCLE

Anxiety is a common cause of not sleeping well or suffering from insomnia. While not the only reason or leading culprit, late-night smartphone usage contributes to the disruption of our sleep cycle.

Getting on the bed and mindlessly scrolling down will keep your brain stimulated with new information when your mind really should calm down for the end of the day.

When we wake up from our phone's alarm, the first thing we do is check our lock screen for the recent updates that came in while we were asleep. We open the opportunity to see stressful emails from our boss, awful news, irrelevant material, or any other stress-inducing stuff. Who wants that when they first start their day?

Is it the right way to begin a fresh day? You should buy an alarm clock and keep your phone in another room. Don't check the device until you've been awake for 20 minutes and finished your morning routine. Doing this will help reset your sleeping pattern.

FROM ESCAPE TO ENVY

As discussed in previous chapters, we sometimes use our smartphones to escape from reality. Smartphones can be a gateway for running away from our real-life troubles by distracting us with entertainment that doesn't matter. Whether to forget about stress or to remove boredom, the smartphone can be a pathway to addiction. We often feel that browsing makes us happy and relieved when watching what others are up to in their lives. Seeing and thinking about these posts compels us to compare ourselves with everyone else. There is a tendency to be filled with subtle envy. Repeatedly, as we browse every day, the little bits of jealousy add up, even if we overlook the individual comparisons we make.

THE LITERAL COST OF A SMARTPHONE

All emotional and physical potential issues aside, some of us are in a financial rat race when it comes to our smartphones. There's invariably hype about the latest models, luxurious brands, sophisticated features, sleek designs, high-definition cameras, and all that. We always keep looking for the newest and the best.

Compared to their earliest costs, smartphones are now available in the various price ranges, from under $100 used to over $1500 brand new.

If you found out your friend bought the new iPhone 12 Pro today (or whatever is the latest and greatest model at the time you read this), you may consider upgrading your phone. Maybe you decide you want it and try to justify why you should upgrade or how to come up with the money to pay for the new model.

This desire is how, without even knowing, you drag yourself into stress and anxiety regarding your smartphone, obsessing about the expense and how to manage it. Depending on your financial situation, you may have even taken out a loan or financing offer from whoever was selling the phone. Even 0% interest financing can cause anxiety, too, because then you have to remember to make another monthly payment!

But think about it for a while. Is this hassle worth it? How long will you keep loving your new iPhone 12 Pro? Only until Apple launches the next iPhone 13 Pro into the market?

CHAPTER TEN

A Smartphone Balanced Life

THE REWARD: A HAPPIER LIFESTYLE

Why do you need to overcome your smartphone addiction? We discussed how smartphone addiction is affecting your livelihood. It is time to understand how you can improve your life as you come to terms with your smartphone addiction. Life is not a bed of roses. So, it is not like you'll suddenly have a perfectly balanced life once you've beaten your smartphone addiction.

However, detachment of your phone encourages you to focus on the other things in your life. You can have a clearer vision of your loved ones and your hobbies. Likewise, when you come out from virtual reality, you can see your real life around you close up.

It also aids you to maintain your physical and mental health. How? Keep reading below.

LESS CHANCE OF ANXIETY AND DEPRESSION

Anxiety and depression are two possible symptoms of cell phone addiction. In some people, it can be severe. Imagine life at a lower risk of developing these two disorders; you can lead a healthy life without having to stress out because of your smartphone. You can have a fresh mind without getting hung up over the posts you see online. You can bask in nature without checking your status. You can find yourself or just enjoy the time or the meal without screen time. Your device can go back to being a tool rather than a hindrance to your mental health. Your eyes will thank you too!

BETTER RELATIONSHIPS

Your loved ones want the authentic you, not the tailored version you post on social media. When you put your smartphone aside and focus on who is next to you in the real world, you will connect with them stronger than you can do by a direct message. It helps to have a better relationship with your friends and family. For those who are living far away, take the moment to call them and check in. It's more personal than a standard "What's up?" text that you could have just copied and pasted to your other contacts at the same time.

REGAINED FOCUS

The frequent use of a cell phone can blur your judgment and affect your ability to concentrate. You can enjoy the clearer focus once you put your phone aside. We covered this in more detail in our chapters about notifications and productivity.

When you are no longer interrupted by notifications at inconvenient times, you can perform without disruption.

Remember that distraction-free time helps you with your tasks, and you'll feel more in tune with what you're doing. So, when you have a stronger focus, you can move further towards success. The key is just to put your phone aside to where you are in control.

MORE HOBBIES

How much time are you putting in on your phone? If you're a smartphone addict, there's a chance you're spending over 5 to 6 hours a day with your cell phone. What if you put those hours into some other hobbies? Isn't there anything you'd rather do if you could put even half of that time toward something else?

Unless you have planned to watch Netflix for these newly freed hours, you have limitless potential!

CHAPTER ELEVEN
Getting Over the FOMO

IT'S NOT EASY

According to a 2015 Virginia Commonwealth report, *Can You Please Put Your Phone Away? Examining how the FOMO phenomenon and mobile phone addiction affect human relationships*[19], compared to other smartphone addiction factors, fear of missing out (FOMO) contributes 75%. In the earlier chapters, we have already discussed how this fear can urge people to check on their notifications more and more in a loop.

To get over your smartphone addiction, you must overcome your FOMO. Remember that once you conquer this fear, winning over smartphone addiction becomes more comfortable. Once you acknowledge this, the rest of the defeating of cell phone addiction is simple.

Be warned that getting over FOMO is not an easy feat.

---

[19] Chaudhry, Laila A., "Can You Please Put Your Phone Away? Examining how the FOMO phenomenon and mobile phone addiction affect human relationships" (2015).
(https://scholarscompass.vcu.edu/uresposters/143)

Recall that the F in FOMO is FEAR. It means you'll require immense willpower to fight your fear. We have already discussed how you can stay motivated and determined to get rid of smartphone addiction. Now you need to overcome the biggest obstacle in the way.

APPLY THE JOMO (JOY OF MISSING OUT)

You might wonder how it is possible to feel joy while feeling like you are missing out. You may feel mixed feelings, as though you are purposely avoiding going to a party that you'd sort of like to attend! So how can there be a peace of mind and satisfaction by skipping it?

There will always be other parties, and you can choose which ones you'll go to in the future. If anything significant takes place at the party you miss, someone will tell you about it afterward anyway.

In the same way, you can skip out on following everybody and everything on social media. If something big happens, such as the next Marvel movie getting announced, you'll soon find out about it. It's impossible not to overhear to the big story nowadays, even if you were to 100% disconnect from the internet. You might not discover about an announcement on Day 1 of the report, but depending on the importance, you'll hear about it shortly.

That said, think of all the news feed crap that you won't have to filter through to learn the same news secondhand. If it turns out you are interested in finding more details, you can always search for it and call it done. Rather than having spent hours reading endless newsfeeds each day, you can research what's really important in 30 seconds. The time you'll save is precious. You need not compete with your friends about who heard it first. You'll gain an enjoyment of life instead of investing your time into a virtual one where most of the content doesn't affect you.

That is the underlying feature of the Joy of Missing Out.

THE REALITY OF THE FILTERED REALITY

One of the significant realities of social media is that it is not reality at all.

Most of the "truth" we see on social media is the filtered reality. People cut out what is ugly and useless and show only what is attractive, attention-seeking, or entertaining. They prefer to hide the dull or unpleasant aspects of their life. Most of your newsfeed stories are refined through rose-tinted glasses. Seeing so much of this clutter can give you a warped sense of reality.

Keeping in mind that what is on social media isn't the only thing that is real. A lot is going on outside of your awareness, and this is the key to not feeling negativity because of everyone else's positivity. You know all about your good parts and bad parts, but you can't always know about anybody else's troubles. People don't always show their struggles on the internet, but they exist.

When you accept that everything you see in social media is filtered truth, you will be on the path to understanding reality. This acceptance allows you to take the next step to overcome FOMO: gratitude.

APPRECIATION THROUGH GRATITUDE

The practice of appreciation helps you to be grateful for the things you have. You can see what you have rather than what others have. Remember that everybody has something and lacks something else. You need to know what you have instead of what other people have.

When you have a smartphone addiction, you need to understand that celebrities' gossip and lifestyles should not be your priority. The most significant act of gratitude is to appreciate your own life and adequately acknowledge it. The latest news, politics, or gossips are not worthy of your time. It would be best to decide what to look at rather than pressured what to see, but you need first to recognize what you have.

This kind of feeling helps you conquer the FOMO. Remember that once you overcome FOMO as an enormous obstacle, you can help yourself get over smartphone addiction quickly. FOMO is the first major step!

CHAPTER TWELVE

Going on a Smartphone Diet

A PLAN FOR SMARTPHONE USAGE

Making a plan for how to beat your smartphone addiction is the first step you can actively take toward balancing your smartphone addiction and reducing your anxiety. There is a legitimate need to check your device every day if you are a working professional or a student. However, rather than taking in all messages and notifications as they come in, you can handle this more efficiently. Setting a schedule when you'll check your phone to answer messages will put yourself at less risk of overwhelming yourself and spontaneously opening non-important apps. For instance, you might schedule some checks at 9:00 am, Noon, 3:00 pm, and the end of your workday.

For instance, you can set a goal that you will only use your smartphone for 5 minutes at a time when you use it. Similarly, you can develop a plan for how often you'll let yourself spontaneously check your phone. For example, you can grant yourself the opportunity to use your phone freely every day during your lunch hour. Experiment with different schedules and find a balance between feeling comfortable and reducing your overall use of your device.

If you're failing with one goal, stick to it or adjust it until you get success. Otherwise, there will be no meaning in setting any goals. In this way, you'll know how much time you have to complete one task.

LET YOUR FRIENDS KNOW

When you are not using the phone, it would be better to keep it out of sight. You should choose a place where you cannot see your device but can still access it if an emergency comes up. "Do Not Disturb" mode is your friend here. Most Apple and Android smartphones have those settings easy to set up.

If you schedule exclusive time slots when you'll be checking in on your phone, you can let some of your closest friends and family know about your plan. It might shock them when you don't instantly respond like you usually would do for them. If you explain that you're trying to reduce your mobile phone usage, but assure them you will reply to their message by the end of the day, they won't worry about you. Communication is the key to not feeling pressured to respond to someone's messages ASAP.

CHAPTER THIRTEEN
Smartphone Minimalism

UNSUBSCRIBE AND UNFOLLOW

When your wardrobe is overflowing with the clothes and accessories that you don't wear anymore, it's an opportunity for decluttering. Decluttering is an art of throwing out the useless things in your life and keeping what you truly need. Unsubscribing and unfollowing pages and people is the most crucial process of decluttering your newsfeed. You don't require the constant update from the website or celebrity that you've been subscribed to for months or years. It's better to check in on them when you may randomly think of them rather than be updated several times a week.

Make a chart and write which channels or individuals you absolutely need to receive versus who is optional. You should try to unsubscribe and unfollow the different channels and newsletters who are not vital to your well-being. Unsubscribing can help you focus on the essential things in your life. You can do this in the following way:

1. Log in to each of your social media accounts and go to your friend list or followed-pages details to see your page subscriptions.
2. Try to sort or assign each person or page into one of these three categories: family, entertainment, or news.
3. Consider everything you learned in this book and work to trim your lists down to only what is essential to follow. For all else, unsubscribe or unfollow it.
4. If you don't wish to remove a person or page from your account entirely, that is okay. Write the name of the page or person in your notes and then unfollow them anyway. If you crave to check on them in a week, use your list and manually go to their page to see their latest updates.

The point of separating all of these pages and accounts out is that it puts you in command. Manually checking pages allows you to view those pages and people whenever you choose to look at them rather than whenever your social media website decides to show you. By unsubscribing, you are now in control over social media websites!

Remember that letting go of the things that seem essential can be hard on you and may feel like you are going to lose something important. This is the FOMO feeling that we discussed in the earlier chapter. If you're reluctant to follow the steps above, take the entire process at a slow pace. For example, you can unfollow and unsubscribe from only 20% of the least important accounts and people. Once you succeed in this task, you can unsubscribe and unfollow another 20% a week later.

Easing into this strategy will ease you out the pain of letting go.

LOG EVERYTHING

Logging your smartphone habits on paper will help you analyze your growth and remember it. The purpose of this is that you won't be able to lie to yourself about how many moments you've spent, or make excuses about why you're using your phone more than you planned. So, it would be best if you wrote your progress for the entire day. Whenever you look at your phone, even for a glance, log it. A lot of phones have this feature built-in already. By manually writing on a sheet of paper, you'll have a higher awareness of the number of times you're checking your device each day and how much time you're spending.

Why use paper when there are already many built-in apps on your phone that can track your screen time? You can use the tracking apps to help you count the total time spent, however this is not enough. Taking the effort to write manually your usage habits down on paper requires more conscious effort to help you acknowledge your smartphone usage. It also helps you mentally separate your smartphone habits from where you are trying to stop it. If you were an alcoholic, you wouldn't go to a bar to try and stop drinking alcohol, right?

Doing the manual logging on paper every day will give you the chance to reflect rather than just swiping away a notification that told you how many minutes you spent on your phone recently or glancing at a screen with some graphs. In my opinion, conscious acknowledgement and reflection is more helpful than any "tracker" app on the market despite the extra effort involved.

HOW TO TURN ON DO NOT DISTURB MODE ON IPHONE (IOS) AND ANDROID

"Do Not Disturb" mode is a useful phone setting for you, and it is also a vital tool for helping yourself manage your smartphone addiction. Both Apple and Google provide the most to date information on how to configure Do Not Disturb mode on their phones. Here are some direct links to their support pages about that in case you do not know how to turn it on:

- Apple (iOS): https://support.apple.com/en-us/HT204321
- Android: https://support.google.com/android/answer/9069335

Regardless of the other steps you've taken so far to combat your smartphone addiction, please schedule the Do Not Disturb mode when possible and gain more control over your smartphone situation.

UNINSTALL/MOVE UNNECESSARY APPS

On your smartphone's home screen, I bet you've sorted your apps into particular places and folders. I also bet you have good muscle memory and can open your favorite apps without looking at your screen. If you discover that you're habitually accessing the same apps repeatedly, consider moving one of those apps to a different folder than usual so you can disrupt your muscle memory. You'll become more aware of the fact that you're opening that app. Moving your apps can be very useful in breaking your patterns. For instance, if you find that whenever you're done visiting Facebook that you always check Instagram next or some other typical behavior. Because you've moved the app out of its regular place, you must consciously swipe into the new folder where the next app is to open it. Doing this will bring you more awareness about your habits. Bonus points if you can uninstall the app and only reinstall it when you're planning to use it. Using this strategy of moving or uninstalling apps will help you a lot with your smartphone journey!

ARE SMARTPHONES WEAKENING YOUR FAMILY RELATIONSHIPS?

Many of us love to spend our day on the smartphone, and in this whole scenario, we forget that we have several friends and family nearby. The fact is that using smartphones regularly will weaken your relationships if you ignore or stop noticing who is in front of you. If you are continually looking at your phone and not interacting with the people around you, this activity can damage your relations at a massive level. Especially if you have kids, this can harm their relationship with you long term.

While smartphones can connect you to the entire world, they also disconnect you from the life in front of you. It is perfectly acceptable for some people to use smartphones while sitting around their family. Still, it is crucial to be mindful of how much time you spend on your phone versus conversing with your family members.

If you prefer to bring the entire family onto the same page, then it is an excellent opportunity to call for a household meeting. Suppose the members of your household use their smartphones around each other and rarely converse face to face. In that case, get everyone involved with setting realistic expectations about smartphone usage.

THE FINAL NOTE ON SMARTPHONE MINIMALISM

Breaking free of your smartphone is hard, and there will be times when you need to browse and mindlessly distract yourself so you can unwind! The difference between using a smartphone effectively and being addicted to it is that it should be a conscious, deliberate action each time you use your smartphone. It's not about the number of apps you have installed; it's about how you manage them and how they affect your mind and lifestyle! Awareness and purpose are key!

CHAPTER FOURTEEN
Summary and Conclusion

YOU'VE LEARNED A LOT!

Smartphones are one of the most sophisticated technologies in the world and serve billions of people every day. Smartphones undoubtedly help for our ease and comfort and provide means for communication, academics, professional excellence, e-commerce platforms, and many other functions.

Just like the coin has two sides, these devices' overuse can contribute to extreme disorders in our world. When you mindlessly use your phone and become too dependent on it, this can lead to smartphone addiction. Although mobile phone addiction is not yet widely considered a psychological disorder, it shares many symptoms. It can negatively affect your way of life, just like a drug addiction does.

Mobile phone addicts suffer from a kind of fear called nomophobia. If you are afraid of being away from your phone or have a compulsion to check your phone regularly, there is a possibility that you have nomophobia.

Wasting a considerable amount of life on your device makes you lose your productivity, creativity, real-world social relations, and mental well-being.

Multiple factors can make you a smartphone addict such as loneliness, anxiety, depression, FOMO (Fear of Missing Out) on the news, or a desire to escape from real life. We get ourselves addicted to our smartphones for many reasons, including but not limited to over stimulation from notifications, attractive offers, clickbait news headlines, and comparing ourselves to others' social media presence. The most important thing to remember is that smartphone users are only a means to earn revenue for the companies behind the applications. Sometimes the benefits and conveniences of a smartphone do not outweigh the adverse outcomes.

We cannot doubt that our smartphones have many features to love, but mindfulness is key to ensuring that we don't go overboard with our usage. Anxiety, procrastination, and loss of productivity are common problems associated with the addictive use of smartphones.

Our smartphones stimulate and affect our brains, similar to pleasure-inducing drugs and other behaviors such as gambling. The constant stimulation destabilizes the natural dopamine levels in our minds. This repeated cycle of smartphone usage gives us happiness at first, but eventually can lead to anxiety and depression.

Online news and media blogs are also guilty of causing these problems. Clickbait article headlines and thumbnails compel us to click on the links we see. To escape this, keep in mind that not all news and the latest updates are worth our acknowledgment. We must learn to differentiate between factual news and what is merely an opinion piece or speculative article of "news."

Becoming a subscription minimalist will help cut down on the overstimulation from smartphone usage. Unfollowing, unsubscribing, and uninstalling time-consuming apps and channels, muting unnecessary notifications, and customizing your smartphone for restricted use can support overcoming smartphone addiction.

Getting rid of the compulsive need of your smartphone will lead to a better and healthier lifestyle for you. Your relationships, learning, professional growth, productivity, and life are not the things you should put at stake for merely a digital gateway.

THANK YOU AND GOOD LUCK

I appreciate you investing the time to read this book, and I hope that most of it has been helpful to you. We are on this journey of navigating the digital world together! Remember that if at first you don't succeed at seizing command over your smartphone, take a few steps back and start over at the basics. I've suffered from smartphone addiction multiple times in my life before I gained control again in 2017. I still use my smartphone every day, but in moderation. My phone rarely disrupts me anymore, and I no longer feel anxious about my social media presence. I believe you can have that too; that's why I wrote this work, and I am thankful you read it.

As an extra thank you to you for buying this book, you can go to the publisher of this book's website and download a free cheat sheet PDF file that you can print out or save in your notes. Look at it frequently to remind you of everything you've read in this book so you can stay focused on what really matters in your life. No sign-up or registration is needed to get the free PDF cheat sheet.

Here's the URL: https://bookpopmedia.com/smartphone-guide

I hope that this book has been helpful to you somehow or that you can use the information in this publication to help someone you love! If you've found this book useful, I'd much appreciate a 5 Star Review on Amazon and any feedback you have for me as the author. I'd love it if you told me your story and whether the ideas and concepts helped you with your smartphone relationship.

Best wishes to you and thank you again!

***

Printed and distributed in the United States of America

https://www.bookpopmedia.com

Made in the USA
Monee, IL
13 September 2026

62976609R00049